SoC Emulation—Bursting Into Its Prime

Bernard Murphy—SemiWiki
Daniel Nenni—SemiWiki
Mentor Emulation Team—Mentor Graphics, a Siemens Business

A Semiwiki.com Project

SoC Emulation—Bursting Into Its Prime

@2017 by Daniel Nenni and Bernard Murphy

All rights reserved. No part of this work covered by the copyright herein may be reproduced, transmitted, stored, or used in any form or by any means graphic, electronic, or mechanical, including but not limited to photocopying, recording, scanning, taping, digitizing, web distribution, information networks, or information storage and retrieval systems, except as permitted under Section 107 or 108 of the 1976 US Copyright Act, without the prior written permission of the publisher.

Published by SemiWiki LLC
Danville, CA

Although the authors and publisher have made every effort to ensure the accuracy and completeness of information contained in this book, we assume no responsibility for errors, inaccuracies, omissions, or any inconsistency herein.

First printing: January 2018
Printed in the United States of America

Edited by:
Mentor Emulation Team—Mentor Graphics, a Siemens Business

Storage Market Case Study:
Ben Whitehead, Storage Product Specialist, Paul Morrison, Solutions Specialist and Shakeel Jeeawoody, Emulation Strategic Alliances, Mentor, a Siemens Business

Contents

Introduction – Why Emulation	1
Verification and Validation	2
The Performance Wall for Simulation	3
The Beginnings of Emulation in Hardware Design	5
Chapter 1 – ICE is Nice	7
Arrays of FPGAs	8
The Good, the Bad and the Not So Bad	10
Emulation Versus FPGA Prototyping	12
Chapter 2 – Three Architectures	14
Processor-Based Emulation	15
Emulation Based on Custom FPGAs	17
Emulation Based on Commercial FPGAs	20
Timing in the Three Architectures	22
Chapter 3 - Emulation Muscles onto Simulation Turf	24
Speeding Up the Test Environment	24
Multi-User Support	26
A Change in Approach to Debug	29
Chapter 4 - Accelerating Software-Based Verification	31
The Rise of Software in Electronics	31
Why Software is Important in Verification	32
Emulation and Hardware/Software Co-verification	33
All Software isn't Equally Important to V&V	34
Debugging in Hardware/Software Co-verification	35
Chapter 5 – Beyond Traditional Verification	36
Performance Modeling	36
Power Modeling	36
Test Verification	37
Deterministic ICE	38
Complex Load Modeling	39
Chapter 6 – The Role of Emulation in Design Today	40
Virtual Prototyping and Emulation	40
Simulation and Emulation	41
Emulation and FPGA Prototyping	44
The Outlook for Emulation	45
Storage Market and Emulation Case Study	46
State of Storage	47
Current leading HDD and SSD Storage Technologies	47
HDD Controllers and Associated Challenges	49

SSD Controllers and Associated Challenges	51
Typical Verification Flow/Methodology	52
Pre-silicon Verification	52
Post-silicon Verification	53
Gaps in Current Methodology	53
Simulation and FPGA Prototyping Methodologies	53
Is the Verification Gap Getting Better?	54
Is increasing Firmware helping current methodologies?	56
Is Hardware Emulation a Viable Option?	56
Hardware Emulation in the flow	56
In-Circuit-Emulation (ICE) Mode	57
Virtual Emulation Mode	61
Implementing an SSD Controller in Mentor Graphics Veloce	64
Creating a Verification or Validation Environment	64
Running the Tests	65
Debugging	66
A/B Testing	66

Conclusion   67

References   69

Introduction – Why Emulation

The modern electronic systems enabling smartphones, smart cars, smart everything are so amazingly capable because we have learned to compress incredible capability into tiny chips the size of a fingernail, more than we were able to pack into a large room not much more than 50 years ago.

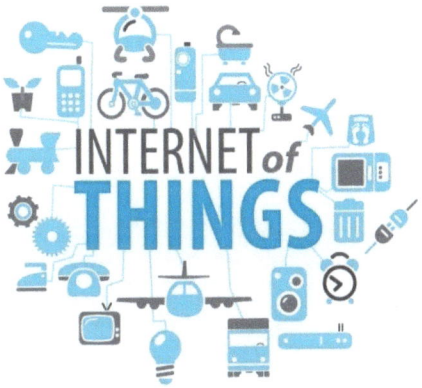

But this amazing capability comes with a cost. Anything we can build, we can build wrong which means we must also test what we build to find (and fix) any potential problems. In the early days, we only had to check that our room-sized computer could do arithmetic on smallish numbers correctly and move data between punched cards or paper tape, simple magnetic-core memory and teletypes. In fairness, there were plenty of other mechanical and electrical reliability problems to address but once solved, testing that the computer functioned correctly was not especially challenging.

Modern semiconductor engineering has largely solved the reliability problems, but now we hold in our hands systems a billion times more functionally complex and therefore considerably harder to test. Now accurate testing has in many ways become vastly more challenging. As consumers, we expect these marvels to function perfectly when making phone calls, browsing the Internet and so on and we'll happily switch to another supplier if we're not satisfied. More importantly we expect these devices to be safe and secure and keep us safe and secure. And all of this must be satisfied in small device, often running on a tiny battery that you might re-charge a few times a week. Test is not an area where we want to hear our device was "reasonably well tested".

Verification and Validation

Testing that these complex chips will function correctly is called verification (did I build what I was told to build) and validation (does it do the job It's supposed to do, often not quite the same thing), which together is usually called V&V. Chip manufacturing has become so incredibly expensive and time-consuming that, to the greatest extent possible, simple economics demand extensive V&V before the chip is built. Which brings us to simulation and emulation.

These days, any complex engineering objective is modeled extensively before it is built, from subsystems (automatic transmissions, fuel injection, jet engines, turbines) to full systems (cars, aircraft, power plants). We must do this because trial-and-error design (build it, then test it, then re-build it and re-test it, …) would be unreasonably expensive and slow. So we build models and "simulate" those models operating under many different conditions of stress, temperature, airflow and other conditions before we commit to manufacturing.

The same applies to chip design; we build a model of the chip and simulate it under many different conditions – user interaction, data traffic, internet activity, video and audio activity and all the other functions that might be active. Fortunately, in chip design the model we create to eventually build the chip is also the model we can use to drive V&V.

Simulation is typically software-based and, for semiconductor design, is a large program, which models the primitive logic functions which make up the design together with the interaction between those functions, responding to models for external and internal data traffic. This "bottom-up" simulation, mimicking the behavior of the full system as the interaction of all those primitive functions and stimuli, has been the mainstay of V&V for digital chips for many decades.

The Performance Wall for Simulation

Software-based simulation has many advantages. At least in theory it's easy to change the simulation model and stimulus; all you must do is change the software description of the design. And it's very easy to see exactly what's going on inside the model, which is important when it comes to debugging why some piece of design functionality isn't working correctly. But there is a significant drawback – it's not very fast and it becomes even slower as the design size increases.

This wasn't so much of a problem when "big" in chip design meant a few million logic gates and some memory. Simulator builders found clever ways to make the software run faster and run many simulations in parallel. But complexity has been growing exponentially in graphics chips, networking chips, the application processors at the heart of smart phones and in many more large systems. There can be a billion or more primitive logic functions in such a device, a scale which would reduce a software simulation model to a crawl.

Unfortunately, that's only one part of the problem. The size and complexity of the testing required for V&V has grown even faster. We now must consider internet and radio traffic, audio and video streams and increasingly complex internal processing in multiple functions at the heart of the chip; these must be modeled with enough accuracy and across enough representative use-cases to give high confidence in the completeness of V&V.

Meanwhile, active power management on these devices has sections of a chip frequently switching between full speed and running slower or turning off, either to conserve battery power or to keep temperature within reasonable bounds. This behavior is controlled by a combination of software and hardware interactions, and therefore you need to simulate both hardware and software together, across all those functional use-cases, to have high confidence in the correctness of the design.

You could imagine that even more cleverness could have kept the game going for a bit longer for software-based simulators, but it was already clear many years ago that a new approach would become essential for large-scale designs and testing. That's where emulation comes in.

SoC Emulation—Bursting Into Its Prime

The Beginnings of Emulation in Hardware Design

Emulation in the larger (non-chip) world generally means making one operating system (or a process on that OS) behave like another, purely as a software process. Software development platforms for mobile phones are (in part) emulators in this sense. In the chip design world, emulation has developed a slightly different meaning; what we want to emulate is a chip (hardware), rather than an OS (software), but otherwise the concepts are similar.

Android emulator on a PC.

But we don't just want to make the emulator behave like the hardware – we also want it to run much faster than a software-based simulator. There's a general principle in the electronics world – if you want to make a software function run faster, you convert it into hardware. In software, you must fetch instructions from memory, execute them and store results back in memory, all constrained to run one stage per clock cycle (in the simplest case). On a chip, you can skip the fetch and store parts (because that's built into the logic model) and greatly simplify the execute part; you can even combine executes from multiple steps into one clock cycle. And you can run as many stages in parallel as practical. In short, hardware can run many orders of magnitude faster than equivalent software, so this should be a good way to accelerate V&V.

But hardware is expensive to build, so you want to build something that will accelerate V&V for a lot of different kinds of design. In 1983 IBM announced they had built the Yorktown Simulation Engine (YSE)[1]. YSE was a custom multi-instruction, multi-data (MIMD) machine designed specifically to replace the TEGAS[2] software simulator. This demonstrated multiple orders of magnitude improvement in performance over TEGAS, validating the principle. A slightly later model, also from IBM and called the Engineering Verification Engine (EVE)[3] further improved on the earlier model. Daisy Systems and Valid Logic, among others, also introduced specialized accelerators around the same time.

Powerful though they were, these systems still suffered from two problems: (a) by building on the architecture of a software simulator, they lost performance they might have had in an implementation more closely mirroring the real logic on the chip and (b) they were still constrained by very low performance in getting stimulus into the hardware and returning the results from the simulation. Both these functions were generally too complex to be accelerated by the same hardware, so still had to run outside the accelerator at the speed of software running on conventional computers. And since the accelerated design model and the software test model needed to communicate and therefore had to synchronize frequently, net gain in performance was much lower than had been hoped.

Chapter 1 – ICE is Nice

To recap, early simulation accelerators suffered from two problems: native simulation speeds were still much slower than real chips and slowed down even further in connecting to external software testbenches.

Emerging in the early to mid-1980s, Field-programmable gate-array (FPGA) technology[4] provided an excellent solution to the first problem. With this technology, circuit netlists up to a certain size could be mapped onto an FPGA in a form not too different from that implemented on a real chip, though interspersed with circuit overhead the device required to support programmability (and re-programmability) to any arbitrary digital circuit. This improvement allowed circuit models to run at speeds in the order of 100's of kHz, still much slower than the ultimate chip being modeled, which would run at many MHz, but much faster than earlier software or accelerated software simulators (~ 10-100Hz at a few million gates[5] and dropping lower as design size increased).

The second problem – slow performance of software-based testbenches and monitoring software – turned out to be an opportunity in disguise. Building software-based tests to accurately model interaction of the design with external devices like disk-drives and networks was becoming increasingly difficult. And testing comprehensively across many possible interactions at these speeds was clearly impossible.

The opportunity was the realization that perhaps it was better to throw out software tests and instead use the real electronics to provide stimulus to the design model, and to consume outputs. You build the real electronic system, but use the FPGA-based model of the design in place of the chip you are currently designing. And since the system is running at hardware speed, there is no slowdown caused by trying to synchronize with verification software. This method of modeling is called In-Circuit Emulation or ICE.

The advantages of ICE are significant:

- You run at speeds much higher than software simulation, so you can cover a much broader range of tests.
- The stimulus and responses you are dealing with are realistic for systems in which the chip will be used, raising confidence in completeness of V&V.
- Asynchronous behavior in the system, which often exposes bugs in a design, is handled naturally.

In fact, for all these reasons, ICE modeling is a very popular use-mode for emulation today. So, verification problem solved? Of course, it's never quite that easy. The FPGA-based solution comes with its own challenges.

Arrays of FPGAs

Early FPGAs could model circuits of several thousands of gates, rather smaller than a typical digital chip design. The technology has advanced rapidly, to the point that devices come pre-integrated with millions of completely custom logic cells, DSP slices, many megabytes of memory, fast transceivers and much more. But digital design continues to grow too, with the result that a large design often must be mapped onto more than one FPGA and those FPGAs must be connected on a board, that board then representing the full model for the design.

This isn't quite as simple as it sounds. You must divide the design into pieces each of which will fit in an FPGA; this is called partitioning. You can't make this decision arbitrarily because (a) signals needing to run from one FPGA to another, through package pins and board traces, will be run much slower than on-chip connections, which is going to mess up the timing unless you divide logic in the right places, and (b) FPGAs have a limited number of package pins so you're also constrained on how many signals can cross between FPGAs.

FPGA pin limitations were and remain a severe problem, particularly around areas of high interconnectivity between function blocks partitioned to different FPGAs. According to an empirical observation known as Rent's rule[6], the more logic you push into a block, the more external connections it will require. This significantly restricted effective utilization of available logic in each FPGA. One way to mitigate this problem, familiar to readers who understand bus-based design concepts, and introduced commercially by Quickturn (subsequently Cadence) was to introduce interconnect chips[7] between the FPGAs; this provided an extra degree of freedom in managing connectivity in partitioning[8].

Another approach called VirtualWire[9], developed by Virtual Machine Works (subsequently acquired by IKOS, which was then acquired by Mentor Graphics), pipelined multiple logical signals from inside an FPGA onto a single package pin, in effect reducing high connectivity between packages through time multiplexing.

There were other challenges, particularly on-chip clock timing where clock routing could be unpredictable, but through FPGA and software technology advances most of these problems have been reduced to manageable levels, if not entirely eliminated. Growing FPGA capacity alone has helped significantly[10].

The Good, the Bad and the Not So Bad

ICE remains a popular use mode for emulation for good reason. This is still a good way to verify functionality in a very realistic environment without needing to first build the chip. That said, ICE comes with its share of challenges beyond those mentioned earlier.

First, when you decide to build a custom chip, you likely do so in part because custom design will enable you to offer the highest-possible performance product. But when it comes to modeling the design, an ICE model, since it is emulating behavior, will inevitably run somewhat slower than at least some of the surrounding devices to which it must connect. Managing this lower performance requires synchronization to adapt the speed of those devices to the emulation model and vice-versa. This is accomplished using speed-bridges, each designed for a specific interface.

For example, if you are designing a component which must connect to a PCI interface, you will need to use a speed bridge to connect your emulation to the other end of that connection – perhaps a microprocessor. The microprocessor will run at full speed; when it sends a request to the emulation model, the bridge will forward the request and will then repeatedly stall the microprocessor with re-try responses until the emulation model is ready to respond. So speed bridges reduce net system performance and, since they are extra hardware, add to the overall cost of the emulation solution.

ICE setups can suffer from reliability problems. Cabling between the emulation socket on the full-system board and the emulator can lead to electrical, EMI and mechanical problems which add debug time to your verification objective. But when a solution is otherwise ideal for your needs, you find solutions to these problems. V&V with realistic traffic can be a very compelling advantage, cables notwithstanding. This is particularly important when software models for external traffic are not available or when you feel they do not sufficiently cover a wide range of real traffic.

One last issue. Initially, ICE setups were single-purpose, single user resources. This was partly a function of limitations in what the emulator itself could do (it was designed as a single-user, single objective machine) and partly because connections, possibly through speed bridges, to a system board were necessarily unique to that setup. Emulators are also expensive. Along with speed bridges, setup and maintenance, this made ICE sufficiently expensive and time-consuming that it was only used only by customers with deep pockets, on the largest and most complex designs.

Emulation Versus FPGA Prototyping

What I have described so far may sound just like FPGA prototyping (see for example the SemiWiki book "Prototypical"[11]). There's a reason for that. On the family tree of verification methods, FPGA prototyping and emulation solutions started in the same place, but then evolved into different solutions, albeit in some cases with similar architectures.

Prototyping has focused, unsurprisingly, on providing the most effective platforms for early system and software development, assuming a quite stable (late-stage) design architecture. Prototyping performance is expected to be as high as possible to support reasonable turn-times for system / software development, so considerable time must be spent setting up and optimizing the prototype before it is ready for use. Emulation on the other hand has focused more on earlier stage chip-design development which requires supporting quick turn-around for design changes and detailed debug.

To give a sense of these differences, an emulation may run at ~ 1-2MHz where a prototype can run at 5-10Mhz or even higher, given sufficient hand-tuning. On the other hand, emulation runs will typically take less than a day to compile, whereas a prototype may take from 1-3 months to setup. Also debug in prototyping systems is typically limited to a small number of pre-determined signals, whereas debug probing in emulators is essentially unlimited.

In short, emulators and prototyping systems today solve different though related problems. Emulation is ideal for verifying and debugging evolving designs, but not for supporting heavy-duty software development. Prototyping is much more helpful to support the application software development task and is general more cost-effective for that task, but is less useful for debugging hardware problems and is not a very effective platform when the chip design is changing rapidly. Vendors who support both provide integrated flows which greatly simplify switching back and forth between these platforms for full system V&V[12].

Chapter 2 – Three Architectures

For in-circuit emulation, emulators were clearly the only game in town, but software simulation continued to dominate the V&V process in all other areas. However, simulators struggled on full-designs and large subsystems, since on relatively small designs (<3M gates) they might reach ~10-100Hz real-time performance, completely impractical for testing systems over millions or billions of cycles, and becoming even slower as design size increased. When verification teams saw what was possible in emulation, it was natural for them to wonder if they also could benefit.

An obvious approach to speed up simulation - to massively parallelize - has demonstrated about an order of magnitude improvement in speed[13] [14]. While impressive and certainly of value for simulation-centric jobs, these solutions only appeared relatively recently and remain too slow for the many big V&V tasks that have become commonplace.

ICE emulation has the raw performance but simulation needs demanded a lot of work on the underlying architecture, which consequently evolved in three directions– processor-based, FPGA-based and custom FPGA-based. Each approach has strengths and of course some weaknesses. We'll look at these below.

Leading-edge emulation hardware includes several components, such as chassis, boards, chips, power supplies, and fans.

Processor-Based Emulation

While emulation started with FPGA-based architectures, this is not the only possible approach. Processor-based architectures have also enjoyed considerable success, but you shouldn't confuse processors in this context with standard CPUs. These processors are custom chips, each of which is an array of small interconnected CPUs. The original concept and early versions were developed by IBM, building on earlier architectures. This was then licensed by Quickturn and introduced commercially in the late 1990's. Cadence developed this architecture into what we now know as the Palladium platform[15].

The basic architecture is quite easy to understand[16]. (I should add that what follows is based on a 2005 reference. I am sure that many features have evolved since then.) Each CPU can evaluate Boolean expressions. When the design is compiled, all expressions are reduced to 4-input forms to simplify the hardware architecture. Evaluation scheduling is determined according to familiar leveling and other algorithms, with an obvious goal to reduce the number of time steps required through as much parallelism as possible. You might think of this approach as a little like native-compiled simulation but supercharged with massive parallelism.

There are additional details in compilation. Combinational loops must be broken (by inserting a flop) and tristate busses must be mapped to two-state "equivalents", both cases being simple to handle. Memories can be mapped and/or synthesized for register banks for more efficient implementation, and interfaces to external hardware (such as devices on an ICE board) must be mapped.

Of course, the whole emulator isn't on one chip. A full system is built from racks of boards, each board including multiple multi-chip modules (MCMs), each MCM carrying multiple die, each of which hosts multiple processors. Part of the job of the compiler is to optimally distribute analysis across these many processors in the total system.

Compilation speed is a strength for this approach, allowing for large (~1B gates) designs to compile from a single host in a day. Industry reports tend to show platforms of this type having an edge in compile times over other approaches.

Run speeds are in the 1-2MHz range which is comparable to the custom FPGA approach, though not as fast as the commercial FPGA approach, which is closer to FPGA prototyping (and therefore typically has slower compile times).

Debug accessibility is also strong. All CPU outputs are accessible and can be stored in a trace-buffer over some period to provide good visibility in debug. Debug data can also be streamed out at high speeds. I should note that debug for all emulator vendors is generally viewed as a post-process (I'll discuss this later in this book).

Emulation Based on Custom FPGAs[17]

Mentor had been experimenting with their own FPGA-based emulator, but in the mid-1990s switched to an architecture based on their acquisition of Meta Systems, who had built a custom reconfigurable device designed specifically to accelerate emulation. I should note that Mentor prefers to call this a custom reconfigurable or reprogrammable device rather than an FPGA because it really it is a special purpose and custom design specifically architected for emulation to overcome many of the disadvantages of general-purpose FPGAs. However, there is no widely-understood name for this type of device so I shall simply refer to it (where appropriate) as a custom FPGA.

In this architecture, primitive logic elements (Lookup Tables or LUTs) are organized with routing in a clever hierarchically self-similar fashion, this being designed, together with specialized place and route software, to minimize congestion in routing between elements and to ensure balanced timing in connections.

I mentioned earlier utilization limits in using conventional FPGAs for emulation. The timing problem arises because FPGA platforms are built with the expectation that the designer will fine-tune timing, at least to some extent, and that they are not (necessarily) concerned with exactly matching the cycle-level timing of a reference design. This is OK when you are prepared to spend days or even weeks optimizing the design, but not when you want a quick and accurate compile to start emulation runs. The Meta Systems architecture addresses both the congestion and balanced timing problems.

These improvements in support of emulation come at a cost, having a somewhat lower LUT count per unit area than a general-purpose FPGA of the size, so it wouldn't be the right solution for general-purpose devices, but it is ideal for this application.

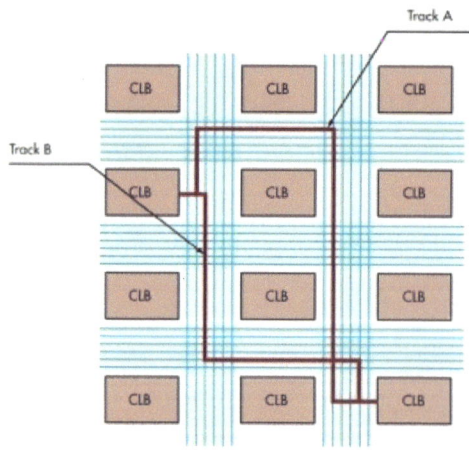

The custom reconfigurable or reprogrammable device improves routing over commercial FPGAs.

The second innovation came from IKOS, acquired by Mentor in 2002. Part of the objective was to timing-multiplex signals between custom FPGAs (to overcome package pin count limitations), but it actually does more than that. It first compiles the design into a giant virtual custom FPGA, then maps it through multiple stages of timing re-synthesis into the array of actual custom FPGAs available on the emulator. At the same time, memory re-synthesis maps memory into devices on the board rather than within the custom FPGAs. From a compile point of view, this whole process is more deterministic than for mapping onto commercial FPGAs, and is typically quite a bit faster than for systems based on such FPGAs. The architecture also leads to higher utilization per custom FPGA, somewhat offsetting the lower device count per unit area.

Thanks to the custom architecture, compilation speed seems in practice to be very good, and in the same general range as for the processor-based approach. As for all architectures, large designs must be partitioned; also after mapping to devices, per-device sub-designs must be physically implemented on each custom FPGA; this task is typically farmed out to parallel computation on multiple PCs.

Run speed is in the same 1-2MHz range as for the processor-based approach, again not as fast as commercial FPGA, which gets its speed (and slower setup times) by being a scaled-back version of FPGA prototyping.

Debug accessibility is also strong since this is built-in to the Meta Systems architecture. Internal node visibility is comparable to the processor-based approach, also stored in a trace-buffer over a selected time-window and debug data can be streamed out at high speeds in support of debug and other applications.

Emulation Based on Commercial FPGAs[18]

Emulation started with this approach, but the most popular platform that continues to use this architecture was introduced first in 2003 by EVE (subsequently acquired by Synopsys and now marketed as the ZeBu server[19]). As with other architectures, a big design won't fit into a single device, which means the design must be partitioned, though commercial FPGAs can achieve similar capacities to other approaches with a smaller number of devices, leading to smaller emulation boxes, with shorter interconnect and therefore faster run-time performance[20]. As for the other architectures, good partitioning is essential to getting good run-times.

When the emulator is built on commercial FPGAs, as is the case in the Synopsys ZeBu system, use of partial crossbar devices mitigates timing-balancing problems due to inter-chip interconnect by controlling interconnectivity between FPGAs though a bus-like structure.

A second complexity is that each piece of logic targeted to a partition must be synthesized, placed and routed within that FPGA. Emulation vendors whose solutions are built on commercial FPGAs provide scripts to run these steps automatically, but the only certain way to ensure push-button simplicity is to aim for sufficiently low utilization per device that the logic in the partition is guaranteed to fit and meet timing constraints without need for manual intervention. This works, but the capacity per FPGA for these systems tends to be lower than you might expect from the advertised capacity of the devices, unless you manually intervene during setup, which can significantly slow compile times.

SoC Emulation—Bursting Into Its Prime

Also, even when push-button, FPGA synthesis, place and route is designed to generate high-quality mapping onto complex underlying architectures, which takes time. (I should note here that Synopsys uses its own software for partitioning and synthesis rather than the FPGA vendor software, so we should expect this step is optimized for their application.) As for custom FPGAs, this mapping for each partition can be implemented separately, so all of them can be run in parallel on multiple PCs. Even with parallelization, this leads to compile times for commercial FPGA-based solutions which can significantly trail those for other architectures.

An advantage of the commercial FPGA approach is run-time performance. Recent industry data seems to indicate run-times as much as 5 times faster than for the other architectures, thanks to more effective gate-packing / utilization per unit area; this in turn allows for implementation on a smaller number of FPGAs, and therefore you can expect smaller interconnect delays between devices since a smaller number of devices can be closer together on a board. If long regression runs dominate the bulk of your testing (where a long setup time is not so important), this can be a real advantage. These systems can also have an advantage in capacity over custom FPGA approaches, though it is not always clear how much effort must be put into compile to reap this benefit[21].

An important downside for commercial FPGA platforms is in debug. Unsurprisingly, FPGA vendors don't assign significant value to debug access to all the internals of a design; their market goals are functionality and performance in the end-application. Some (SRAM-based) FPGAs have a path to access every node, primarily for bitstream programming, but this is generally too slow for debug purposes. Alternatively, it is possible to add debug paths to any node which are then compiled into the design, but these decisions must be made before compilation and can have noticeable impact on emulation performance. For these reasons, debuggability remains a weaker point for systems based on commercial FPGA devices.

Timing in the Three Architectures

Almost all digital design is synchronous, which means that logic evaluations starting at a given time are constrained to complete in a clock cycle so the results of those evaluations are all ready when you want to start the next round of evaluations. While most designs today use multiple clocks, each part of a design is controlled by just one of those clocks at any given time. Correct operation of the whole design is very dependent on carefully balancing logic evaluation against clock cycles and carefully managing the distribution of clocks around the chip, so that clock latency (delays in arrival of a clock at any given part of the circuit) is balanced across the whole device.

In emulation, the same clock balancing must apply not just within each emulator device but across the whole system, requiring careful management of clock generation and distribution through boards, racks and even between cabinets when large enough designs must be modeled

When it comes to signal timing, it might at first seem that FPGA architectures would have an advantage over processor architectures because the FPGA designs look more like real circuits. But after mapping design gates into lookup tables and mapping clocks onto the pre-designed clock distribution system of an FPGA, resemblance between the FPGA implementation and the ultimate design implementation is limited. Additionally, each architecture effectively models the multiple clock frequencies in the real design based on modifications to a common clock for the emulator. So any apparent advantage one architecture may have over another in this area is largely in the eye of the beholder.

A second potential problem in timing can arise for data signals split across partitions. Even when optimal partitions are found, that does not guarantee that high-speed signals will not need to cross between chips, board or cabinets. Added delays in such signals, through package pins and board-level traces between chips, may force a need for synchronization to keep other signals in lock-step, which can dramatically drag down effective performance for the emulator.

Managing system-level delays effectively can be challenging since interconnect delays at the chip, board and rack levels can be dramatically different. Minimizing these differences requires use of the same interconnect technologies common in advanced server platforms, such as optical and Infiniband interfaces[22]. At this level, emulation technology is starting to share advances most often seen in advanced datacenters.

Chapter 3 - Emulation Muscles onto Simulation Turf

Architectural advances in emulation paved the path for full-chip and large-subsystem V&V engineers to seriously consider emulation in addressing their modeling needs. The primary problems they needed to have addressed to consider emulation as a worthy alternative to simulation were:

- Make setup a push-button process
- Make the stimulus-generation process run at or near the speed of the emulated design
- Provide debug access comparable to that offered by a software simulator, with minimal impact on performance
- Make the solution multi-user (given an appropriate license, many users can launch simulation runs at the same time; the same should be possible for emulation)

The setup and debug needs were handled through differing approaches in the 3 architectures. Accelerating the test environment and making these systems truly multi-user became the next major areas of development.

Speeding Up the Test Environment

The model for the design, typically known as the device under test (DUT), emulates quickly in principle but faced significant slowdown when communicating with a software-based testbench. Those tests run on a conventional computer and, since they typically communicated with the DUT through hundreds of signals and these were required to synchronize between the (comparatively slow) software test and the emulated DUT model, emulator speed could be dragged down close to software simulation speeds. ICE solved this problem by connecting the emulation model directly to real hardware, but that approach would often be too difficult to setup and inflexible for simulation users (though exceptions are starting to be supported in methods called "In-Circuit Simulation Acceleration").

Another solution was to move the testbench to the emulator, in which case everything was running at emulation speed. Sometimes called targetless acceleration, this approach is frequently a part of the solution, but it isn't a universal solution. What goes on an emulator must be mapped into low-level functions and therefore must be synthesizable, which limits the complexity of stimulus generation and of assertions, when thinking of debug. Put another way, you move testbench components to the emulator that you could equally implement in hardware. Another consideration may be that modern testbenches can be significantly larger than the DUT (in lines of code). Even if they could be ported to the emulator, cost to emulate the full testbench plus DUT could be prohibitive.

Still, targetless acceleration covers a useful subset of testing needs – synthesizable IP, synthesizable verification IP and synthesizable assertions (such as OVL assertions) are obvious examples. But loops with complex bounds, complex assertions, class-based structures and software-based tests – these must stay outside the emulator.

In this same vein, emulation vendors now support common verification IP, with some limitations, on the emulation platform; this is accomplished using proven design IP equivalents in synthesizable model form. Since a lot of V&V, for example performance testing, will typically replace many IP with VIP models, this is a very effective way to move major chunks of functionality to the emulation platform[23].

Another breakthrough, first introduced by IKOS, is transaction-based interfacing. Instead of communicating between testbench and emulator through bit- or word-level signals, this method communicates through transactions or protocol packets as bundled groups of signals[24]. In the testbench running on a PC, a high-level model (in C/C++, SystemVerilog or similar languages) can send and receive these abstracted packets directly. This function on the host computer is mirrored on the emulator by a synthesizable bus functional model which can convert packets to signals and vice-versa, all at emulation speeds. The real benefit here comes from both sides being able to stream transaction packets as they become ready rather than stalling the emulator for each signal/word update; this greatly reduces impact on overall performance.

Multi-User Support

All these improvements were critically important to make emulation effective in simulation use-modes but emulator unit prices (in the millions of dollars) would have severely limited adoption if systems remained single-user, as they were in the very early days of emulation. Fully aware of that limitation, IKOS (subsequently acquired by Mentor in 2001) offered multi-user support by 1998[25], within a similar timeframe Quickturn (subsequently acquired by Cadence in 1998) announced support for multiple users[26]. Eve (subsequently acquired by Synopsys in 2012), announced support in 2009[27].

But this capability was limited. Usage in ICE mode still tied up a machine and even when not running in ICE mode, multi-user still didn't look like the virtual capability that we have come to expect from public and private cloud services with support for remote access, queuing, round-the-clock utilization and scalable growth in capacity. Stepping up to these requirements was essential to become an effective alternative to software-based simulation which could already run on scalable server farms and private clouds.

Cadence[28] and Mentor[29] now support true virtualization platforms to meet these needs. Users can login from any location to launch jobs and download results for debug. These systems support job prioritization, queuing and scheduling and aim to maximize utilization and throughput, just as you would expect for conventional jobs in a server/cloud environment.

Since ICE remains an important use-mode, it must continue to be supported in virtualization. This is accomplished through at least a couple of methods. Cadence provides an emulation development kit[30] which is a rack-mountable system providing extension to external devices through any number of ports. Through this interface, you can add external devices to your virtualized emulation. The connectivity to external devices is also virtualized – external devices in this kind of setup can be shared between different jobs running on the emulator.

SoC Emulation—Bursting Into Its Prime

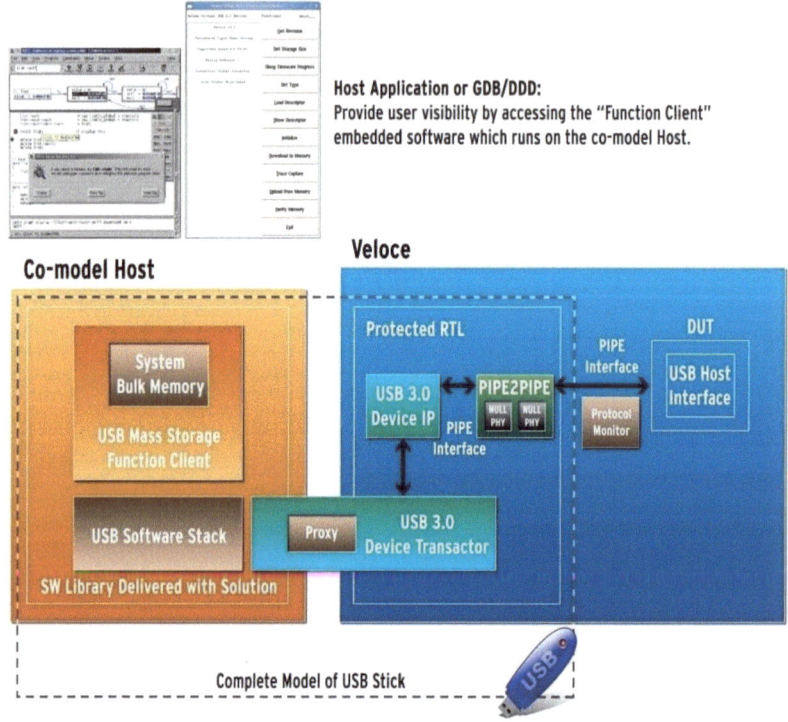

Mentor VirtuaLAB co-modeling.

Mentor VirtuaLAB[31] offers a software-based solution, still connecting to real traffic but here through the host system. VirtuaLAB splits the peripheral between a design IP running on the emulator and corresponding software stack/application running on the host. Since everything is virtualized in this model, again you can run ICE modeling from anywhere in the world.

There's debate about the pros and cons of hardware-based and software-based modeling of external devices; each seems to have merits in differing contexts. This is a topic I have touched on in a SemiWiki blog[32].

A Change in Approach to Debug[33]

The great thing about emulators is that they run very quickly. But they aren't optimal for interactive debug. In debug, you want to set triggers and breakpoints. When you hit a breakpoint, you want to poke around to understand why that trigger tripped, then you want to trace back to figure out what caused those events until you hopefully get back to a root-cause for the problem.

But this takes time. A verification engineer doesn't think at MHz speeds and even if they could, all that searching around until they find the "ah-ha!" bug takes a lot of discovery. You really don't want to have an emulator spinning its very expensive wheels while the engineer is doing all this thinking. So, by far the most common approach to debug in emulation is to dump all the debug output to a big file, then do offline analysis in a (software-based) debugger, which doesn't consume utilization dollars from your department budget as you search.

This raises the obvious question of "how do I know what to dump, or when to dump it, if I don't yet know I have a bug?" The answer comes in two parts: assertions and trace-buffers.

Assertion-based verification has been common for some time in software development yet started to appear in hardware design more recently, particularly thanks to formal verification methods. The basic idea is quite simple. Rather than waiting for something to go wrong then trying to figure out why, you create multiple assertions about things you believe should always be true at certain critical places in the design. For example, you might create an assertion that a certain buffer should never overflow. You know that if it does, design behavior may be bad, or at least unpredictable.

These assertions can be written in a form which can be directly compiled into the emulator. And they can be (and should be) scattered all over the design, each asserting some expected behavior in critical areas. A design sporting assertions like these is a powerful aid to V&V. In emulation, these assertions typically have negligible impact on emulation performance; they simply stand to the side waiting for the possibility that they might be triggered.

On such a trigger (and/or at times of your choosing) trace buffers in the emulator provide state information for past time and can be dumped to support subsequent debug. Of course, there are bounds on what you can trace and for how long, but these seem to be quite generous in modern emulators. You can also checkpoint and, if you find you hadn't traced quite enough signals to support your debug exploration, you can restart emulation from a checkpoint, avoiding the need to restart from scratch.

The result of all this is that again, instead of waiting for something bad to happen then trying to figure out why (which is very difficult if you can't do interactive debug), you preload the design with assertions on expected behavior. If any of them trigger, you already have a head-start on a much earlier root-cause for that bad thing that would eventually have happened, and you have big debug traces to dig back into why the assertion failed. All of which you can do in offline debug. Incidentally, this isn't just the right way to do debug in an emulator – it's a better way to do debug in general.

Chapter 4 - Accelerating Software-Based Verification

The Rise of Software in Electronics[34]

Back in the mists of time, chips were either computers, such as the microprocessors (uPs) at the heart of your PC or microcontrollers (uCs, a scaled down version of a microprocessor) or they were something more dedicated and hardwired. All the intelligence, driven by software, was in the uP/uC and all the specialized work (perhaps controlling a disk drive or a terminal) was handled by chips needing little or no software. In design, modelling interaction with software was limited to those few uP/uC devices.

But that all changed with the emergence of embedded computers, particularly those from ARM[35]. Small processors sitting on chip can supervise complex functionality and moving data around, allowing complex systems to be integrated onto a single chip where previously they would be implemented with separate components on a PC board. Since this integration dramatically reduces the cost and size of the system, increases reliability and performance and reduces power consumption, access to these capabilities created a stampede towards integration with increasingly sophisticated software control.

But now modelling software with the hardware cannot be limited to testing a few uP/uC devices. Virtually every large chip contains at least one, often multiple embedded computers so every one of those chip designs must be verified with software.

Why Software is Important in Verification[36]

When we start to think of new electronic products, we don't think in terms of hardware or software. We think instead about the functionality we want and how we can maximize the appeal and profitability of that product by maximizing usability and minimizing constraints and cost.

Doing this effectively requires careful tradeoffs between what is implemented in software versus hardware. Some common principles have emerged in SoC design to guide these choices. For example, hardware is faster than software, so you want to implement lots of functional blocks in hardware. But software is preferable when you need to support evolving functionality - apps obviously, but also software in support of embedded needs, such as managing the protocol part of communication. Sometimes you need both. In managing power consumption in a complex product like a smartphone, some power management is controlled by requests from the operating system and some is controlled directly by the hardware. But what ultimately matters for the product, and what requires V&V, is the functionality of the total system. So here's one reason you have to verify software and hardware together.

Another consideration is the way we manage flows of data around the chip. Data can flow in and out from USB, memory and touch screen interfaces and within the device from CPUs, GPUs, DSPs and other functions. As you are looking at a video on the web, listening to an MP3 and connecting to a computer through a USB connection, these functions are competing for attention; this could lead to severe performance problems or even crashes if your design is unprepared to handle these levels of concurrent traffic.

But it would be impossible to design for any possible level of load on the system, independent of software and external data activity. A realistic design must make tradeoffs, so you have to verify against whatever you consider would be reasonable loads. Which again means you must verify the hardware together with software.

Emulation and Hardware/Software Co-verification

Handling co-verification is one place that software-based simulation must surrender. The software part of a test could in principle run on a virtual model on a host computer, communicating with the rest of the emulation model through a transaction-based interface. But performance would be completely unworkable. These tests often need to run millions to billions of cycles. At an effective speed of say 10Hz, even a million cycles would take more than a day. Booting an OS could take years!

Emulation is a much more workable solution (also with the software part of the test running on a host computer), as has been demonstrated by Linux, Android and iOS boots in a matter of hours. Of course, FPGA-based prototyping will be even faster, but emulation is often preferred when hardware debug access is most important.

While emulation is a good solution here, it too has limits and that means care must be taken when considering what software should be included in the emulation-based testing.

All Software isn't Equally Important to V&V

In an ideal world you would run all software, from low-level drivers which provide low-level control of hardware functions, up through the operating system (OS, perhaps Linux), the user interface (often Android or iOS) to all the applications that might run on the device (mapping, music, phone, email, gaming, ..). But this level of testing would make for impractically long run times in emulation. In practice, given agreed interfaces and software testing at the application level, these apps and even lower-levels of software often don't have to be included in hardware V&V.

It is practical to boot an OS and even the user interface in emulation, both of which are important for validating the interaction between hardware and software. For example, Linux and Android can boot in a small number of hours. This can be a little slow for heavy-duty regression testing so it might (late in design) get an assist from FPGA prototyping with a transition to emulation for more detailed debug, or in earlier stages run in "big-job" regressions only after all other tests have passed.

But there's a lot of important software that can be tested frequently in regressions, generally known as "bare-metal" software which is very closely coupled to the hardware. This includes firmware and drivers to connect to each of the addressable functions in the device, such as the GPU and the USB interface as well as support functions like debug and power management. This may be complemented by stripped-down variants of the OS and user interface, with added instrumentation to support more detailed V&V in areas of interest to this design.

Debugging in Hardware/Software Co-verification[37][38][39]

Debug in software-driven verification necessarily becomes more complex. Now you need to be able to chase problems through both software and hardware to identify root causes. All the emulation providers have invested significantly in providing combined hardware and software debugger capabilities to simplify this analysis. This includes methods to view simulation waveforms and transactions, along with memory, registers and stacks, ability to set breakpoints and triggers and all the other features you would expect in both hardware and software debuggers. The days of debug using monitor statements and simulation logs are far behind us - well-designed debuggers have become fundamental to effective debug in hardware/software co-verification.

Chapter 5 – Beyond Traditional Verification

Once you have fast emulation for functional modeling, new opportunities emerge. Some of these are extensions of analyses which are possible, if much slower, with traditional simulation. Others are capabilities which would have been impractical in simulation but with much higher performance become feasible. We'll explore a few here.

Performance Modeling[40][41]

The goal in performance modeling is to determine that the system performs well across a wide range of system use-cases and particularly under heavy loads competing for resources. Doing this sort of analysis is feasible to a limited extent using software-based simulation but requires that many of the components in the design be modeled by verification IP or bus-functional models to reduce the size of the simulation. The danger in that approach is that use of abstracted models may hide potential problems in interaction between fully implemented IPs. Verifying performance with a full model in emulation reduces the chance that you'll miss unusual implementation interactions which may drag performance down in corner-case usage.

Power Modeling[42][43][44]

Power has become at least as important a product differentiator as performance and, in many cases, has become more important. In earlier times, you could put best/worst/typical power values for each block together in a spreadsheet, along with crude approximations of typical use-cases to get an overall estimate of power consumption. But that doesn't cut it anymore. One of the problems with this approach is that it can only give insight into average power. Peak power is also very important. This has an impact on localized heating and voltage drop in power rails, both of which can cause nominally safe timing paths to fail. Peak power can also reduce reliability through electromigration in insufficiently robust power (or signal) routing.

The ideal way to model power is to do so dynamically, so you can see both averages and peaks over a wide variety of use cases. Again, emulation is the best way to do this, but there's a wrinkle. Power must be estimated by summing switching and interconnect power for each node toggle in the design, together with leakage power for each "on" node in the design, where the scaling factors for each of these are pulled from library files. The calculation is not complex, but it needs to be performed across each node on each clock tick, requiring state access to each node in the emulation on each cycle, which would, in a crude implementation, slow emulation significantly.

In all solutions I have seen, power is computed offline rather than in the emulator (all those floating-point multiply/add computations would be far too slow to couple directly to emulation). Vendors who support these flows have streaming methods to output the required data to this computation, while minimizing impact on emulation performance.

Test Verification[45]

Design for Test (DFT) in modern design has become a very complex part of the total design. You have scan test, built-in self-test (BIST) and test decompression logic. This very complex logic, woven in among all the other logic in the design, must be verified, just as you must verify the normal (mission) functional mode of the design. Cycling through scan testing, compression and the complex and lengthy sequences implicit in BIST methods has already become unreasonably time-consuming through software-based simulation, as indicated by the growing popularity of static approaches to test-logic verification. But static methods can only provide limited coverage. Dynamic verification is still required, just as it is for functional verification; emulation can accelerate this objective by factors of thousands or more, making functional verification of DFT logic a realistic option even for large designs.

Deterministic ICE[46]

The value of ICE is in dealing with realistic traffic but that traffic isn't necessarily deterministic, so if you find a bug, it may be difficult to find a way to trace back to pin down the root cause. Running in a mode which also captures traffic provides a way to enable deterministic replay.

In this mode, emulation internal state and external interface input are captured during the course of a normal ICE run. Deterministic ICE support then makes it easy to replay a run based on the captured data. On a replay, you always see exactly the same data seen in the real ICE run so debug is predictable, whereas simply rerunning the ICE run may not reproduce a bug if it was data or environment dependent. You get all the advantages of ICE in getting to see these rare problems, with all the advantages of determinism in debug to isolate the root cause of a problem.

Complex Load Modeling[47]

There are cases where you really need to model realistic loads but even ICE modeling would be too complex. When you're building a big network switch with 128 or more ports and you need to model realistic voice, video, data and wireless traffic in multiple protocols (and software-defined networking) at variable bandwidths over all those ports, setting up an ICE environment would be, if perhaps not impossible, at least extremely expensive and challenging.

In fact, testing big switches in realistic environments is such a big problem that companies have emerged to provide software solutions to model those environments for the express purpose of testing networking solutions. There are several companies in this class (known as network emulation software providers). One such company is Ixia; Mentor has partnered with Ixia to connect their IxNetwork Virtual Edition, through the Veloce Virtual Network App, to Veloce Emulation[48].

Chapter 6 – The Role of Emulation in Design Today

From time to time, debate surges on whether simulation's days are over, to be replaced by a combination of emulation and static/formal analysis[49]. The topic is popular because it ignites entertaining debates around the pros and cons of different methods of modeling. But industry feedback consistently supports a view that multiple different tools for verification are essential and are likely to remain so for the foreseeable future.

It might seem that we are too easily accepting a confusing mix of tools and methodologies. Surely if we could reduce this set to one or at most two tools, costs, training and overall efficiency in V&V could be optimized? In fact, different needs at different stages of architecture and design seem impossible to reconcile into one or two tools. It is worth understanding why, and why emulation providers work hard to provide seamless interoperability between these different tools and flows in support of V&V.

Virtual Prototyping and Emulation

Virtual prototyping may be new to some readers. Imperas[50] provides one popular option; EDA vendors also have offerings in this space. Toolsets like this are most commonly used for embedded software design and development when the hardware platform is not yet available. They start with instruction-accurate models of the underlying hardware, sufficiently accurate that software running on top of the model cannot tell it is not running on the real system, but the model is heavily abstracted to support running at high performance.

Frequently these systems use Just-In-Time (JIT) methods to model when executing the software load, which is why they can't afford to model much detail in the hardware.

Virtual prototypes can run OS and application software at near real-time speed, which is obviously much more effective for software development than the ~1-2MHz speed typical of emulation, or even the speeds offered by FPGA prototyping. And virtual models can be ready for use very early in design planning, unlike FPGA prototypes. But since these prototypes have very limited understanding of detailed hardware architecture, they provide little useful feedback on how the real hardware model will interact with the software.

A hybrid model can bridge the gap[51]. One application is to accelerate adaptation of earlier generations of firmware and OS to new hardware, though it is not always clear how widely this early prototyping use-model is being adopted in practice, largely because software teams often lack cycles to work on planning for the next design when they're busy wrapping up work for the last design. A much more actively-used approach is to support software-based testing of hardware, where software stacks for embedded CPUs run on the virtual model, linked to an emulation of the rest of the hardware.

Simulation and Emulation

It might seem natural that simulation should eventually be replaced by emulation. After all, each is predominantly valuable during design implementation (between architecture design and tapeout) and emulation runs orders of magnitude faster than simulation. But industry veterans in V&V think differently. They see these solutions having complementary strengths, only some of which of which can be consolidated into one solution.

Let's start with performance. Emulation has a huge advantage in run-time, which makes it essential for modeling complete SoC-scale designs and for running software-based verification. Simulation, even accelerated simulation, cannot compete at this level. Speed is also very important when regression testing over significant banks of compliance / compatibility test-suites. This need is particularly common for microprocessor, GPU and other similar systems. And emulation can connect to real-world devices (through speed-bridges) in ICE-mode, a level of verification accuracy that is typically difficult for simulation.

But emulation does not model Z (high-impedance) or X (unknown) states. Modeling these states is often important to correctly analyze bi-directional logic and tristate busses, or to detect state ambiguities which may arise from inadequate reset logic. Emulators are based on 2-state logic (0 and 1), just like real circuits; extending emulation to handle these states would demand significantly greater hardware capacity, making cost very unattractive.

Emulators also can't model detailed timing, which remains popular for some analyses, particularly in checking some aspects of gate-level timing. Another significant limitation is that emulation cannot handle mixed-signal (analog and digital) modeling. Since most large SoCs contain some level of analog interfacing, mixed-signal simulation will continue to be essential in some very important cases. One example is the need to verify training mechanisms for DDR interfaces.

Additionally, simulation is still viewed as easier to use, more flexible and, in the early to mid-stages of design, more cost-effective than emulation. This is certainly true for small blocks but also applies in full system modeling when taking a widely-used layered approach to V&V. Compiling for emulation takes a while (a day for a large design) so it's inefficient to use that solution to find bugs you could find and correct more quickly in simulation. Basic issues such as incorrectly connected clocks and resets, incorrectly set address maps in the bus, incorrectly connected request and grant signals – these can be found quickly in simulation. That leaves emulation to do what it does best - finding the hard, subtle bugs that come up in complex interactions between the software, hardware and realistic external devices.

Some platforms provide methods to "hot-swap" back and forth between simulation and emulation so you can use the speed of emulation to get to important areas for test, then use the simulator to dive down into more flexible debug.

Given these complementary strengths, an increasingly popular idea is to leverage both in a mixed approach commonly called simulation acceleration or in-circuit acceleration, where simulation acts as the master and emulation as a slave to accelerate or provide more realistic modeling through ICE interfaces for some component(s) in the design[52][53]. The performance difference between these systems must be managed, typically through a transaction-based modeling technique, such as SCE-MI[54]. Here instead of communicating signal changes between the two platforms, you bundle and communicate multi-cycle transactions[55], an option now supported on most simulation and emulation platforms.

43

I should add that I have also heard rumblings of co-modeling with mixed-signal simulation, further underlining the value of these hybrid approaches.

Emulation and FPGA Prototyping

FPGA prototyping is a late stage option for design. Even in the best cases, you can be looking at several weeks to a month to setup a prototype, not something you want to be doing when the design is evolving rapidly. A prototype of this kind is primarily valuable in the relatively late stages of system and software development, and primarily for software validation rather than for detailed hardware verification.

For V&V of the hardware design (especially driven by software testing), emulation will dominate, thanks to faster compile times and superior internal visibility for debug. Emulation is also a multi-user capability, especially in virtualization configurations, which is a must-have to support large verification teams, whereas FPGA prototyping is intrinsically limited to a single user at any one time, so necessarily has more restricted usage.

But there's a very useful hybrid mode of operation where an FPGA prototype can be used to get quickly to an interesting point to start detailed hardware debug. From this point a snapshot of memory and other important state can be transferred to an emulator from which analysis can continue with the greater internal visibility needed for that debug.

The value of this approach is based on a common observation - most of the important areas to test (in the hardware) are found after you've got through setup/boot, especially problems in performance and unexpected usage corner behaviors. So a fast, low-visibility path to get to a useful place to debug is not much of a compromise.

Indeed, throughout verification flows, hybrid operation as described in this chapter is becoming an essential way to transition between more abstract and more detailed models, between software-driven and simulation-driven and between slower and faster.

The Outlook for Emulation

Clearly emulation isn't going to obsolete other forms of verification, but it doesn't need to. Adoption continues to grow, as has become apparent in the growth of reported revenues for this segment[56] and in customer pressure on vendors to support virtual operation. Growth seems to be predominantly in mid- to late-stage hardware design, which fits with expectations that reasonably stable designs are not so dependent on ultra-fast compile times and that you can get more value out of multiple verification runs before needing to load an updated/fixed design.

Hardware emulation is here to stay. It seems reasonable to expect continued advances in capacity, performance and capability and increasingly strong and transparent links, not only to other tools in the chip verification flow, but also to more system development and verification tools, as Mentor has already demonstrated with their link to Ixia. Other verification methods will continue to be important in a complete V&V flow for hardware design and close interoperability between all these methods will be increasingly important to streamline verification.

Storage Market and Emulation Case Study

Shakeel Jeeawoody, Emulation Strategic Alliances, Mentor Graphics

For most of us, it is hard to imagine a world without storage capabilities. Storage devices are a must-have for retaining business-critical documents and/or family pictures. Over the years, storage medium has evolved from magnetic tape to floppy disk, hard disk, CD, DAT, DVD and Compact Flash. In fact, the storage evolution is still happening and the pace of innovation just keeps getting faster.

The evolution of data storage.

A better grasp of the dynamics of this market is gained by looking at how the technology has evolved and its growing use. Initially, storage was used for backups. Now it is, to a large degree, moving towards frequent and unstructured data storage, processing and retrieval. This results in a completely new class of performance, reliability and capacity requirements. Additionally, the skyrocketing costs of managing storage systems that comply with legislations for secure information handling has increased the services business model.

The market demands that huge amounts of data/information be stored securely and with accessibility anywhere and anytime. This requirement is driving the adoption of key technologies and use models. The capacity, size and performance of Solid State Drives (SSDs) is making it a very interesting technology for future use. Moreover, the cloud is enabling storage to be more convenient and easier. Non-Volatile Memory Express (NVMe) over Ethernet or fiber channel is becoming a leading solution for connecting appliances to servers.

Current research into new mediums continues with holographic, memristor or even DNA storage. Scientists can already encode text, images, video and even operating systems in DNA. It unlocks enormous potential for video streaming while retaining compactness and durability. The researchers believe that DNA will not degrade over time nor become obsolete.

Having said that, we are far from having these advanced technologies in production.

The question is how to solve today's challenges with the existing set of tools.

State of Storage

Current leading HDD and SSD Storage Technologies

According to GSMAintelligence.com, newly created digital data is doubling every two years. This means increasing amounts of storage must be available at the same pace. As reported by

Statista, Hard Disk Drives (HDDs) continue to be a dominant source of bits shipped, but SSDs are on the growth curve. However, hardware challenges for both HDD and SSD are substantial – and increasing.

Manufacturing hard drives demands a very high investment, in the hundreds of millions of dollars, for clean rooms, ultra-precise robotics and experienced employees. At one time, there were dozens of companies competing. Now there are three: Seagate, Toshiba and Western Digital. Entering this business is prohibitively expensive, and for many reasons, the future of the market is difficult to predict. In addition, all of the patents are in the hands of current manufacturers.

Creating an SSD has fewer financial barriers to entry, by comparison, to manufacturing a magnetic hard disk drive. Many SSD companies combine flash chips with a controller of their own design or one acquired from an outside company, an achievable business model for a wide variety of companies with limited resources. The key differentiators are not the media (flash), which is available from several sources (Intel/Micron, Samsung, Toshiba/Western Digital, SK Hynix and Powerchip).

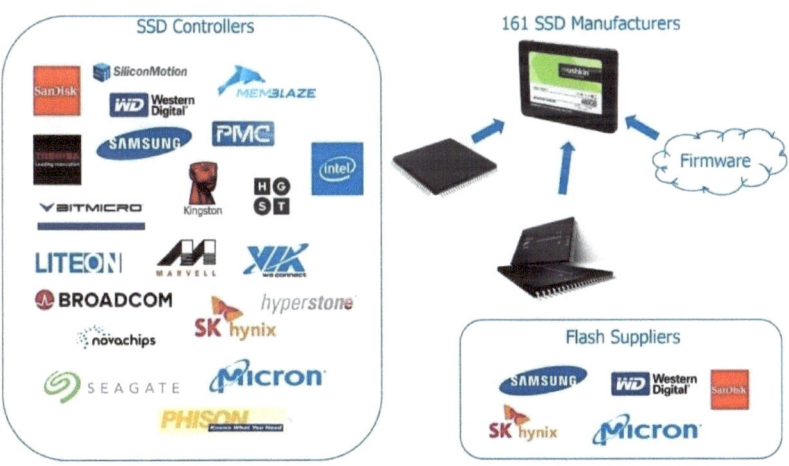

SSD controller market – key players. Courtesy: Mentor

The lynchpin component is the **controller**. Each controller requires an algorithm and firmware (FW) to manage the complexities of writing and reading the various types of flash. This media is changing rapidly: NAND, 3DNAND, 3DXPoint, and other future technologies.

Competition is fierce with the battleground looking much more like the battleground in the HDD industry prior to the consolidation resulting in the big three.

HDD Controllers and Associated Verification Challenges

Hard disk controllers are complex in their own way, with mixed-signal electronics, the usual power and performance constraints and difficulty integrating and debugging 3rd IP, such as the write/read channel and digital back-end with the preamp and servo mechanism.

SoC Emulation—Bursting Into Its Prime

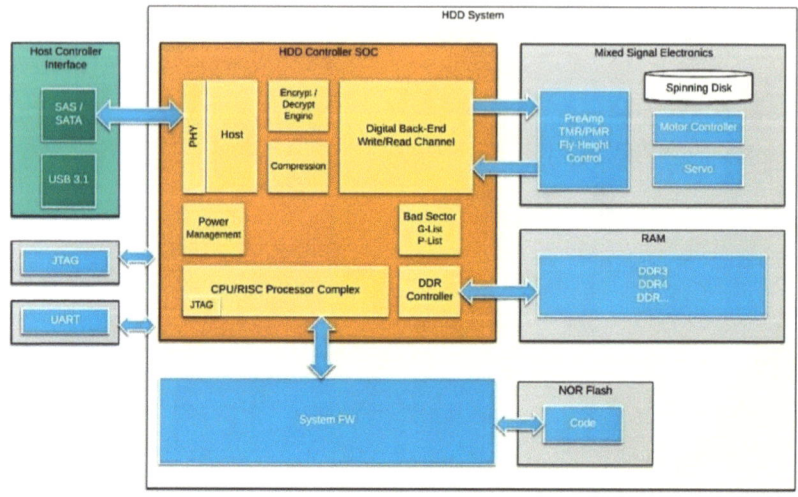

Typical HHD controller SoC verification considerations and challenges. Courtesy: Mentor

Verification engineers must ensure that the complete system works together, with firmware, to claim that a design is verified (see illustration below).

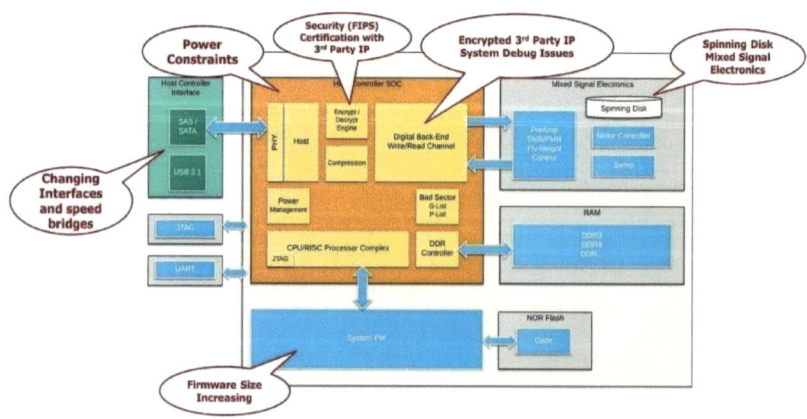

Hard disk controllers are complex with mixed-signal electronics, the usual power and performance constraints, and difficulty integrating and debugging third party IP. Courtesy: Mentor.

SSD Controllers and Associated Verification Challenges

As the industry moves to SSD, the controller faces a surprisingly high number of completely different challenges. Performance of the back-end NAND channels can now saturate a PCIe bus. This was never the case with a spinning disk. It subsequently requires accurate architectural modeling to ensure that power and performance trade-off decisions meet requirements.

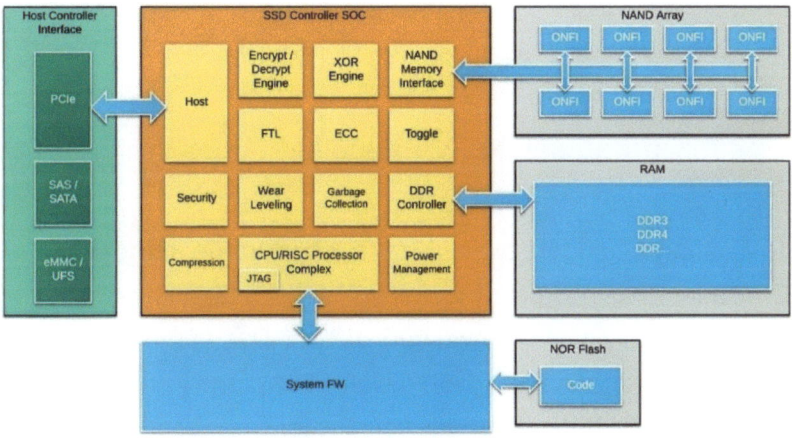

SSD controllers add complexities all their own that need to be managed. Courtesy: Mentor

Managing NAND, in all the various types, requires complex wear leveling, table-management and garbage collection, in addition to all the interface requirements of a hard drive—security, compression and error correction code (ECC).

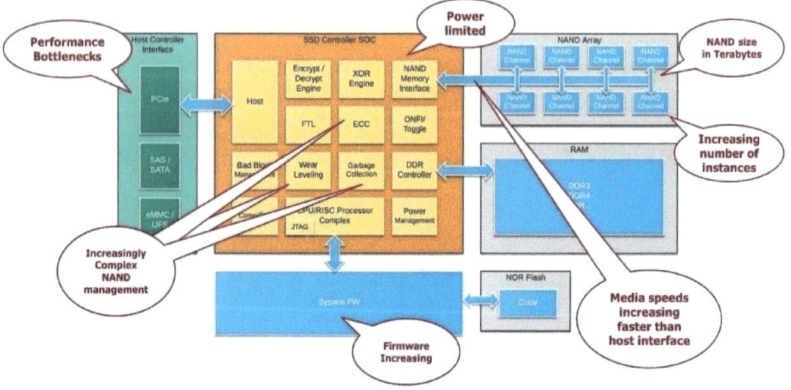

System verification complexity has hit the wall. Courtesy: Mentor

Typical Verification Flow/Methodology

Pre-silicon Verification

Verification engineers typically take a bottom-up approach. They sequentially create block level, sub-module level, module level and finally system-level verification. This approach, at least to the module level, works well when done in-house where teams are located near each other.

As designs get bigger and complexity grows, pieces of the design are purchased from 3rd party vendors. This leads to inevitable integration issues. Furthermore, system-level simulation is no longer a viable alternative as it takes too long, and as such, is not effective.

In some cases, FPGA prototyping for hardware verification and early software development is a tempting solution, but it means time needs to be budgeted to get to a working board. FPGA prototyping, with some care, can ensure proper functionality; however, schedule delays are possible due to partitioning and debugging issues that can turn into a nightmare. Too often, the verification that should be done early enough to enable design changes and trade-off decisions is pushed later in the schedule where FPGA can act as a catchall. The later a bug is identified, the more expensive it is to fix. Finding an architectural problem too late can be a project killer.

Post-silicon Verification

When the chip is back, engineers usually test the design based on the type of adaptor being used—could be PCIe, SATA or SAS.

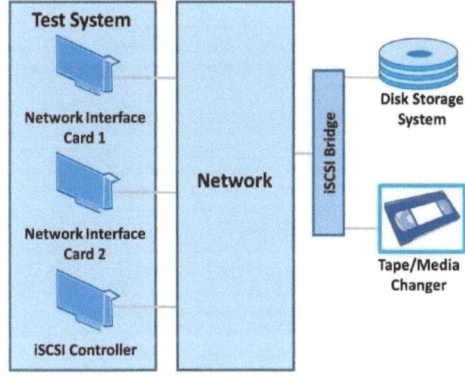

Testing scenario in post-silicon phase. Courtesy: Mentor

They eventually connect to the end-customer library for final validation.

Gaps in Current Methodology

Simulation and FPGA Prototyping Methodologies

Simulation has full visibility into design registers and nets, but as design size increases, it becomes prohibitive to do full system-level simulation. Verification engineers note that it takes too long, is tough to create corner cases in a reasonable timeframe due to lack of stimulus and is difficult to integrate FW into the design and have it behave as expected.

On the other hand, FPGA prototyping, when used for validation, can support full system-level simulation. Unfortunately, FPGA prototyping has limited visibility, making it very hard to debug design issues. Additionally, the FPGA board connects to external hosts, but cannot run at full speed. The FPGA prototype is good for testing a backed-up datapath (due to running slow), but it cannot verify full-speed connections.

Additionally, there are no easy ways to measure real performance until the system-on-chip (SoC) is in the complete system with real firmware. In this case, verification engineers can estimate performance, but habitually miss something.

To check for optimal characteristics based on the shipping configurations, engineers choose to implement A/B testing or split testing. This means running verification for different NAND, different size drives, different configurations and different connectivity. This is possible, but challenging with FPGA prototypes, and close to impossible with simulation.

Is the Verification Gap Getting Better?

Is the verification gap getting better? Actually no, but what causes this?

The nature of flash technologies create some interesting challenges that need to be managed *hours* after the drive is first powered up, after one or more drive-fills.

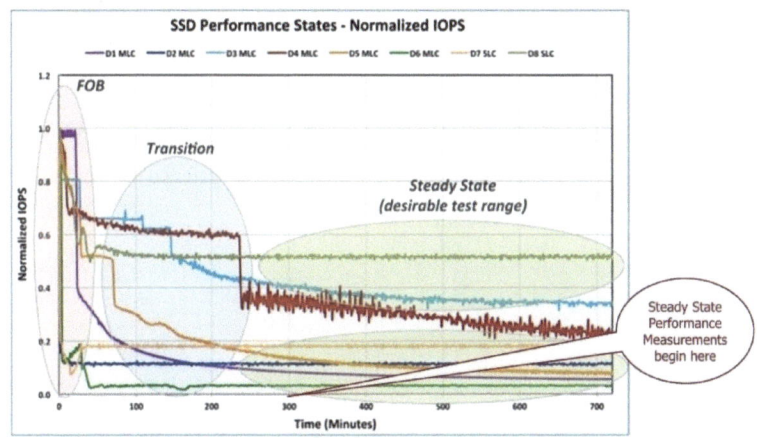

NAND-based SSS Performance states for 8 devices (RND 4KiB writes). Courtesy: Mentor

This new reality of drive performance makes simulating a complete system nearly impossible with traditional methods. It is usually only done for the first time with actual drive hardware and firmware, or with models that pre-load a possible drive-state that create interesting test cases. This can create some unpleasant surprises the first time the drive integration is done.

To solve this, it is important to do performance testing and A/B configuration testing as early as possible. This determines if the proposed architecture and design lives up to its promise.

The option to measure the SSD controller's ability to do garbage collection, while concurrently writing and reading, gives a better indication of real-world performance in comparison to the usual watered down estimation methods of most system-level SSD pre-silicon tests.

Power optimization, security and compression are even more important than ever because the need for secure storage,

SoC Emulation—Bursting Into Its Prime

using less power, is imperative in the data center.

Is increasing Firmware (FW) helping current methodologies?

As firmware increases in size and scope, FW development and verification needs to start earlier, typically happening concurrently with HW design.

Firmware increasing in size and scope. Courtesy: Mentor

Verification engineers are adapting to this complexity by changing the flow:

1) Start FW development concurrent with HW development to find bugs prior to tapeout.

2) Create a plan to test FW with the actual HW (in FPGA or emulation) prior to tapeout, to speed up both FW and HW development and testing. Assume both will be in development at one time, and plan accordingly for debug.

3) Plan for a certain level of design maturity and system-level testing prior to tapeout, otherwise one WILL spin the chip.

Is Hardware Emulation a Viable Option?

Hardware Emulation in the flow

Clearly, hardware emulation is a viable option for many parts of the verification flow.

SoC Emulation—Bursting Into Its Prime

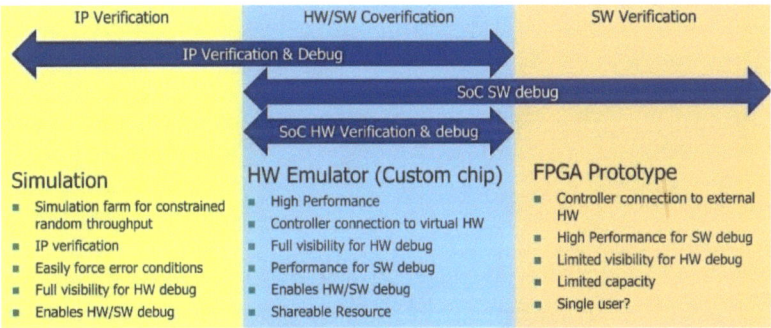

Three foundations of storage verification. Courtesy: Mentor

In general, simulation gives full visibility, lets engineers easily force error conditions, verify design blocks and provides visibility into bugs found in the lab. On the other hand, FPGA prototyping allows for faster and more extensive testing, allows controller connections to external hardware used in or by the drive, and allows for firmware test development with real hardware, but has limited visibility for debug and is much less flexible.

Emulation spans the gap between simulation and FPGA prototyping, as it's faster than simulation, provides more visibility than the FPGA, allows controller connections to external hardware used in or by the drive, runs on real firmware, creates confidence before the FPGA prototype is available, and enables the same setup for both pre- and post-silicon verification. In addition, hardware emulation works in concert with FPGA prototyping for more effective debug and root-cause analysis.

In-Circuit-Emulation (ICE) Mode

In ICE mode, an emulator connects with physical interface devices that allow the execution and debug of embedded systems.

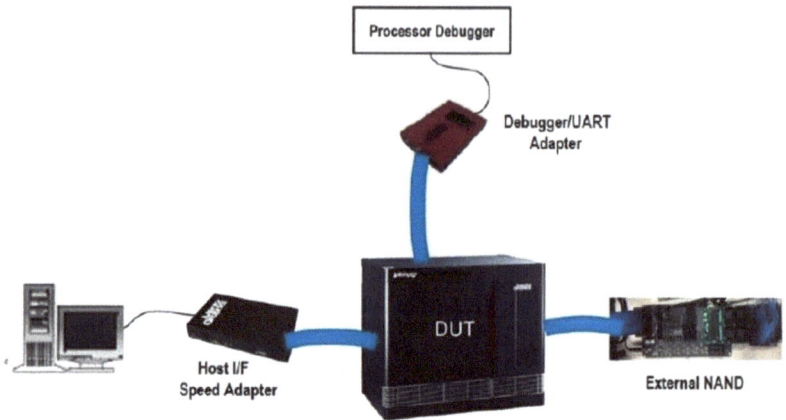

Caption placeholder. Courtesy: Mentor

The debugger adaptor enables processor debug for firmware (rather than just a model) and use validation. The host speed adapter (e.g. PCIe, SATA, and SAS) allows a connection to host testers while reusing existing test scripts. This setup enables designers to develop, test and debug the full test suite against the SoC. Verification engineers find SoC bugs prior to tapeout, and reduce the number of tapeouts required to ship product.

In ICE-mode, an emulator setup for a storage controller might look like this:

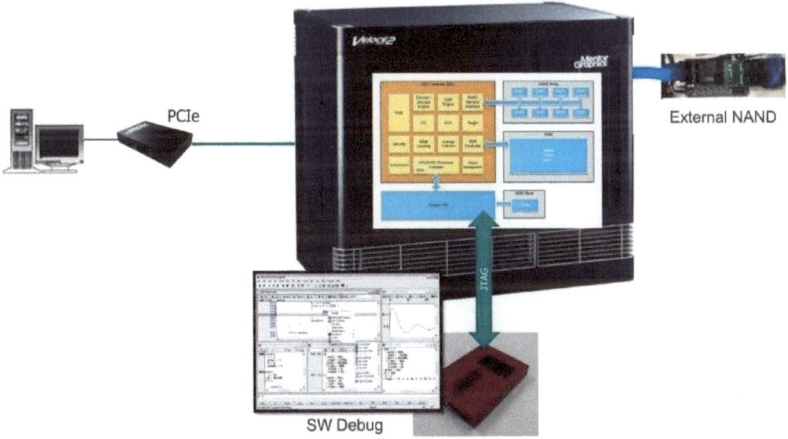

Emulation setup for a storage controller.

There is physical cabling from the emulator to external hardware, in this case the Veloce iSolve PCIe speed adapter, which translates real-world host traffic into emulator speeds. An external daughter-card populated with the NAND devices targeted with the controller is connected, allowing testing of the latest NAND flash. There are NOR and DRAM memory models connected, and a physical JTAG allowing for software debug and step-through of FW code. The bulk of the design under test (DUT) is in RTL and still contained within the emulator. With an ICE approach, similar to FPGA, the system being verified no longer is a one-to-one representation of the final system, as the emulator cannot run as fast as the peripherals connected to it.

Some limitations of this approach should be obvious. We have to create a daughter card for the NAND device, and that device needs to be available.

Clocking the system is now a complex process, and stopping clocks to external devices can have negative and unintended consequences. Many times these consequences keep the external devices from operating properly.

Limitations of speed adapters prevents running the host and the device at the same speed, possibly hiding some timing issues. It also makes it extremely difficult to inject errors, as the speed-adapter does not pass most errors generated from the host to the device.

However, the major limitation comes down to the pain of configuring the system to target different media. A single SSD controller is designed to support multiple SSD capacities via adding or removing NAND channels, or selecting different NAND chips, which will have different number of chip selects, planes, blocks, and possibly even pages within a block and page sizes. Enabling testing support for all of the identified configurations requires, at a minimum, an external card with socketed support for different NAND chips, plus someone to physically swap out NAND chips when switching to a different configuration. This testing is difficult if not impossible to automate, slowing down testing of different configurations. Additionally, controller development is more likely a single project set into a roadmap of multiple controllers, where supporting the current controller and planning for the next is key. That requires support for multiple generations of NAND flash, which further complicates ICE mode testing.

The last big problem is that firmware has become a major part of the functionality delivered in an SSD. The whole system needs to work together to qualify as a product, so testing of just the controller without the associated firmware leaves big holes in the verification and validation methodology. It is clear that firmware has become the focus of SSD success, and has also taken the lead in number of engineers required and schedule time. The hardware and firmware schedules can no longer be serialized if a project is to be competitive. Instead, firmware development needs to start when the hardware is still lacking maturity, and both will need to be debugged together as a system.

Fortunately, there is a way to address each of these problems, which brings us to a virtual environment.

Virtual Emulation Mode

In general, a virtual solution replaces a physical implementation with a model. A purely virtual solution only uses models—no cabling from the emulator to a physical device.

Some companies have had notable success using an approach that virtualizes the parts of the controller that are well understood at an interface level, and allows for much greater flexibility in making design and architectural changes. All while providing greater visibility into the DUT.

If the host design is PCIe/NVMe, for example, the interface itself is standardized and well-known. If we could stimulate that interface in system simulation with something configurable enough to hit corner cases, but simple enough to make bring-up doable in a very short time-frame (without writing an entire testbench to re-invent the wheel) then that would cover a major portion of the controller testing.

At the same time, the NAND interfaces (both Toggle and ONFI) are well-known, but the underlying NAND 3-D technology and device physics are highly complex, and probably still under development if your controller is forward looking. That means the target device probably does not even exist, and there is only an early specification. However, if a model exists for that device, the same process done on the host interface can be done with the NAND interface. Simply drop in a model replacement.

Remembering the quote by George Box, "All models are wrong, some are useful." By understanding that the model does not represent the device hardware exactly, the question remains, is the model good enough? To answer this question, some empirical data is useful. One company started their production firmware development at the same time as the hardware. They used Veloce soft models to emulate the NAND devices, and they found that the firmware that passed on the emulator [using soft models for NAND] had first-pass success when run on the real chip. By design, the DUT on the emulator is identical to the chip.

The fact is, a well-designed model speeds development time and moves firmware integration very early in the project such that any model differences with the actual NAND device, or physical host, is trivial compared to time lost implementing a physical ICE-based system.

Illustrating the "now" virtualized environment, we have a new picture of the system:

SoC Emulation—Bursting Into Its Prime

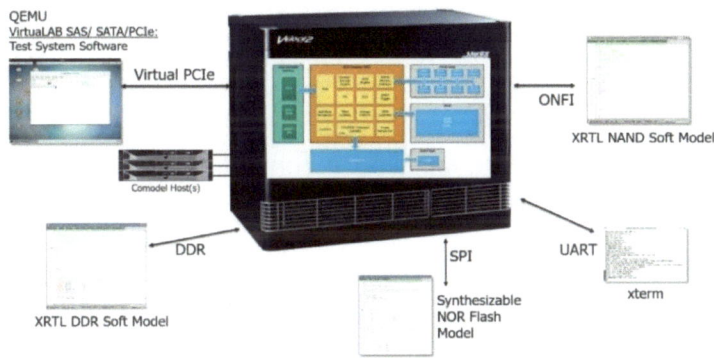

Emulation Deployment on Veloce2. Courtesy: Mentor

Using virtual mode and testing a virtual system also allows the emulator to be part of a data-center use model, enabling engineers to run simulations from their desk and share the emulator with multiple users at the same time.

Firmware development in a virtual environment can start at the same time as design creation. Traditional storage firmware development and testing starts in earnest when the silicon is in the lab, but successful companies have proven software-driven design flows let firmware start with hardware definition. When this happens, the overall design time and the time-to-market shrinks.

The greatest advantage to virtualizing your verification is flexibility. Hard drive controller development rarely attempted targeting multiple variants of spinning media; however, that is exactly what SSD controllers must do. The ability to re-configure a design to target a completely new NAND device, and get accurate performance data, prior to silicon, gives controller teams the advantage.

Implementing an SSD Controller in Veloce™ (Mentor Graphics' Hardware Emulator)

Creating a Verification or Validation Environment

Several steps need to be followed to convert an existing environment for testing an SSD controller on Veloce (see diagram in 4.c above), or creating a new environment for enhanced testing.

The host interface, most likely, requires the most modification to get an environment to run. It is best to use a host VirtuaLAB solution, which connects into a QEMU environment and allows the user to run existing applications that may already be available. This can include test scripts, performance measurement applications, plus any other host-related exerciser scripts. It is highly recommended to use existing test scripts that are part of most regression suites that any SSD drive developer has, that were created for previous products. Users can also run off-the-shelf performance measurement software to measure the performance and identify bottlenecks within the design. All of this creates a better first tapeout and reduces the likelihood or number of subsequent controller spins necessary.

Host interface VTL designs are also available, if emulation of the SSD controller is being used to enhance an existing verification environment. These are similar to existing VIP used within the verification flow, although typically they have a subset of features needed to communicate with a synthesized design.

Replacing the NAND memory with a model is the next highest priority. Given that SSDs typically are sized from several hundred gigabytes to several terabytes, finding enough physical memory available to implement the full drive memory is challenging to say the least. We recommend using a FlexMem memory, a model that runs on a server connected to the emulator with dynamic allocation of the NAND memory as it is used, along with a cache on the emulator hardware to speed up performance. Also available are sparse and full hardware memory models, although those each have restrictions not found in the FlexMem version.

If the environment is going to be used to validate the entire design including FW, a virtual JTAG host needs to be connected to the processor for debug and trace support. We recommend running Codelink® as well, to support quick FW debug.

DRAM and NOR memories also need to be replaced with models. Since these two memory implementations are typically much smaller than the NAND array of memory, HW models that live on the emulator are best used. Also available are DRAM DFI models, which should connect to many DRAM controllers and remove the implementation and debug time required to get a working PHY into the emulator while being guaranteed to work with the Mentor-supplied DDR DRAM model.

Running the Tests

Once the design is ported to the emulator, the user can run a set of tests to check out the controller design. Many testcases should already exist, from the verification environment (used with a VTL host front end) to customer-based validation tests that can check out the design in a full system (used with a VirtuaLAB host front end). While not as fast as real hardware in the lab, these tests will run significantly faster than a verification environment. Many tests NOT even considered before because of runtime, are now possible, running in a fraction of the time while still providing full visibility. FW can also be loaded and run on the emulator, allowing for testing of a production design long before that design is ever available in the lab. This environment supports development and debug of HW designs, FW designs, validation test scripts and customer test scripts all prior to tapeout, reducing time-to-market as well as increasing the likelihood of working first-pass silicon.

Debugging

The emulator has multiple test methods to speed finding the root cause of bugs. While all signals within a design can be captured in a waveform, Veloce also supports a capture mode where only those signals of interest are captured, speeding up run time and finding the bug sooner. Codelink provides a way for FW engineers to run a test, capture the results, and then replay the test while stepping forward and backward to isolate and fix a bug. Capture and replay is also supported from a HW perspective, where the Veloce captures test results for a test run quickly on the Veloce, downloads them to a server to re-run and debug the results while freeing up the emulator for other uses.

A/B Testing

Testing of different SSD configurations, specifically the amount and configuration of NAND memory connected to the controller can be challenging in a typical lab environment. At a minimum, the existing memory is replaced with the new configuration. In the worst case, a new PCB created and parts soldered on the board. This assumes that the physical parts exist and work for cutting-edge development. It's possible that the NAND chips are being developed concurrently with the controller design, and they aren't even available for prototype testing. The Veloce NAND memory model solves those problems. Even if a NAND chip is not available, a specification typically is. A NAND model is created based on that specification and used for pre-tapeout testing of a controller to ensure that it works as expected. If a feature in the NAND chip changes, the model is easily updated to match the new feature and the testing is re-run.

Most if not all controllers are designed to support multiple sizes and configurations of memory, including number of channels, number and size of blocks and pages, number of planes, plus multiple other configuration options. Testing all of these possible configurations is much easier with an emulator. Instead of having to replace chips and possibly create new printed circuit boards, a different top-level file is created that instantiates the different configuration, re-compiled, and a new set of tests run with the new configuration. This makes A/B testing with different configurations and different optimizations much easier and faster to run, allowing the controller design team to make tradeoffs in their design and update their architecture much sooner if it's discovered that there is an unexpected hole in performance or support.

Conclusion

SSDs are fundamentally different from traditional spinning hard drives, and a verification methodology must evolve with the unique challenges of using NAND flash as a storage medium. These differences also reveal new opportunities to use more flexible and powerful tools, including using a virtual host machine driving PCIe traffic, and an entire NAND configuration with the flexibility of soft-models. This frees up the emulator to run multiple users in parallel, creating efficiencies not possible using the ICE mode. Moreover, Veloce's save-and-restore capability is another feature that designers appreciate since it allows them to free the emulator while they debug a previous run.

As storage technology and use models continue to evolve, so do the verification tools needed to solve today's challenges. The Veloce emulator is well suited to address these challenges as experienced by leading storage companies who are using it today in production environments.

References

1 http://mesl.ucsd.edu/gupta/cse291-fpga/Readings/YSE86.pdf
2 http://semiengineering.com/kc/knowledge_center/A-brief-history-of-logic-simulation/12
3 https://www.computer.org/csdl/proceedings/dac/1988/0864/00/00014761.pdf
4 http://www.cis.upenn.edu/~lee/06cse480/lec-fpga.pdf
5 eg. http://www.deepchip.com/items/0522-02.html (graph near end of article)
6 https://en.wikipedia.org/wiki/Rent%27s_rule
7 http://ramp.eecs.berkeley.edu/Publications/RAMP2010_MButts20 Aug (Slides, 8-25-2010).pptx
8 https://verificationacademy.com/verification-horizons/march-2015-volume-11-issue-1/Hardware-Emulation-Three-Decades-of-Evolution
9 http://www.uccs.edu/~gtumbush/4211/Logic Emulation with Virtual Wires.pdf
10 http://www.deepchip.com/items/0522-01.html
11 https://www.amazon.com/Prototypical-Emergence-FPGA-Based-Prototyping-Design/dp/1533391610/ref=sr_1_1?ie=UTF8&qid=1463942374&sr=8-1&keywords=prototypical
12 https://www.semiwiki.com/forum/content/5740-software-driven-verification-drives-tight-links-between-emulation-prototyping.html
13 https://www.cadence.com/content/cadence-www/global/en_US/home/company/newsroom/press-releases/pr-ir/2016/cadence-completes-acquisition-of-rocketick-technologies.html
14 https://www.synopsys.com/cgi-bin/verification/dsdla/pdfr1.cgi?file=vcs-fgp-wp.pdf
15 https://www.cadence.com/content/cadence-www/global/en_US/home/tools/system-design-and-verification/acceleration-and-emulation/palladium-z1.html

16 http://www.eetasia.com/ARTICLES/2005NOV/B/2005NOV01_PL_EDA_TA.pdf?SOURCES=DOWNLOAD
17 http://electronicdesign.com/fpgas/what-s-difference-between-fpga-and-custom-silicon-emulators
18 http://electronicdesign.com/fpgas/what-s-difference-between-fpga-and-custom-silicon-emulators
19 http://www.synopsys.com/Tools/Verification/hardware-verification/emulation/Pages/zebu-server-asic-emulator.aspx
20 https://verificationacademy.com/verification-horizons/november-2015-volume-11-issue-3/hardware-emulation-three-decades-of-evolution-part-iii (Eve/Synopsys section)
21 https://verificationacademy.com/verification-horizons/november-2015-volume-11-issue-3/hardware-emulation-three-decades-of-evolution-part-iii
22 https://www.semiwiki.com/forum/content/5198-r-evolution-hardware-based-simulation-acceleration.html
23 https://www.mentor.com/company/news/mentor-adds-arm-amba-5-ahb-verification-ip, http://ip.cadence.com/ipportfolio/verification-ip/accelerated-vip
24 http://electronicdesign.com/eda/transaction-based-verification-and-emulation-combine-multi-megahertz-verification-performance
25 http://www.eetimes.com/document.asp?doc_id=1212471
26 http://www.thefreelibrary.com/Quickturn+Announces+Palladium,+the+Most+Advanced+Simulation...-a075609988
27 http://embedded-computing.com/news/eves-offers-multi-user-capability-2/
28 https://www.cadence.com/content/cadence-www/global/en_US/home/tools/system-design-and-verification/acceleration-and-emulation/palladium-z1.html?CMP=pr111615_PalladiumZ1
29 http://verificationhorizons.verificationacademy.com/volume-8_issue-2/articles/stream/virtualization-delivers-total-verification-soc-hardware-software-interfaces_vh-v8-i2.pdf
30 https://www.cadence.com/content/dam/cadence-www/global/en_US/documents/tools/system-design-verification/palladium-emulation-development-kit-ds.pdf
31 https://www.mentor.com/products/fv/emulation-systems/virtual-devices

32 https://www.semiwiki.com/forum/content/6711-rise-transaction-based-emulation.html
33 http://s3.mentor.com/public_documents/whitepaper/resources/mentorpaper_81009.pdf
34 http://www.deepchip.com/items/0522-01.html
35 "Mobile Unleashed", Daniel Nenni and Don Dingee, SemiWiki, December 2015.
36 http://embedded-computing.com/guest-blogs/hardware-and-software-grow-ever-closer/
37 https://www.mentor.com/products/fv/codelink/
38 https://www.cadence.com/content/cadence-www/global/en_US/home/tools/system-design-and-verification/software-driven-verification/indago-embedded-sw-debug-app.html
39 https://www.synopsys.com/Tools/Verification/debug/Pages/verdi-hw-sw-ds.aspx
40 http://www.eetimes.com/author.asp?section_id=36&doc_id=1330092
41 https://verificationacademy.com/verification-horizons/november-2015-volume-11-issue-3/hardware-emulation-three-decades-of-evolution-part-iii
42 https://www.cadence.com/content/cadence-www/global/en_US/home/tools/system-design-and-verification/acceleration-and-emulation/palladium-dynamic-power.html
43 https://www.mentor.com/products/fv/emulation-systems/veloce-power-application
44 http://www.synopsys.com/Tools/Verification/hardware-verification/emulation/Pages/zebu-server-asic-emulator.aspx
45 http://semimd.com/blog/2016/02/25/design-for-testability-dft-verified-with-hardware-emulation/
46 http://www.electronicsweekly.com/news/app-based-emulators-go-beyond-rtl-verification-2016-06/
47 https://www.semiwiki.com/forum/content/5742-ecosystem-partnership-effective-network-hardware-design.html
48 https://www.mentor.com/products/fv/techpubs/download?id=97874&contactid=1&PC=L&c=2016_09_05_veloce_ixia_de-risk_network_wp

49 https://dvcon.org/sites/dvcon.org/files/files/2016/Panel-Emulation-Static-Verification-Will-Replace-Simulation.mp3
50 http://www.imperas.com/why-use-virtual-platforms
51 https://www.cadence.com/content/dam/cadence-www/global/en_US/documents/tools/system-design-verification/palladium-xp-ii-wp.pdf
52 https://community.cadence.com/cadence_blogs_8/b/sd/archive/2012/05/16/debug-breakthroughs-enabled-by-in-circuit-acceleration
53 https://www.mentor.com/products/fv/resources/overview/testbench-considerations-for-maximizing-the-speed-of-simulation-acceleration-with-a-hardware-emulator-5b79adaa-0634-41c2-8533-ac88fe5df86b
54 http://accellera.org/downloads/standards/sce-mi
55 https://www.semiwiki.com/forum/content/6711-rise-transaction-based-emulation.html
56 https://dvcon.org/sites/dvcon.org/files/images/2016/DVCon-2016-FINAL%20handout-min.pdf

www.ingramcontent.com/pod-product-compliance
Lightning Source LLC
Chambersburg PA
CBHW040227220526
45473CB00001B/147